靚餸

Tasty Dishes

自煮飯堂

Home-made Cafe

目錄

Contents

牛 · 雞 Beef · Chicken

雞蛋 · Egg

海產 · Seafood

材料介紹 Cooking Ingredients

白蘿蔔
Radish

白蘿蔔可解毒、清積滯、改善吃過量肉類引致的食慾不振。不過白蘿蔔不可與補品同時食用，否則會解除補品的功效。

It can relieve toxin, cure indigestion. Loss of appetite caused by eating too much meat, it can increase appetite. It cannot be served with tonic, otherwise the efficacy of tonic will be reduced.

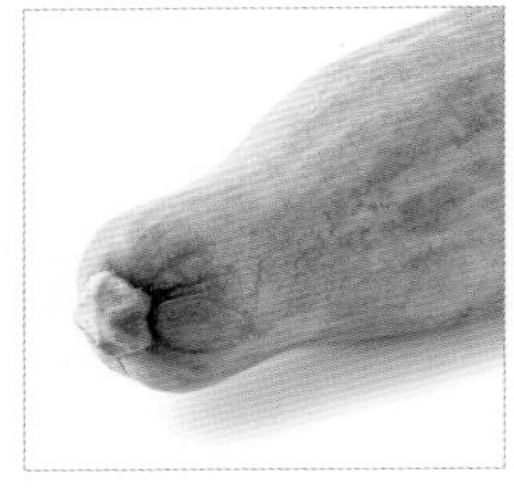

南瓜
Pumpkin

日本南瓜的味道較香甜，口感也較幼滑，但價錢遠高於本地南瓜，讀者可因應需求自行選擇。

Japanese pumpkin has more sweet taste and smooth texture. But the price is relatively higher than the local ones. Depend on what you need, you can choose by yourself.

青瓜
Cucumber

青瓜不宜煮得太久，否則會釋出水分，失去爽脆的口感，更會影響整道菜的味道。

It cannot be cooked for a long time, otherwise water will be released. This affects the crispy texture and the whole dish.

西芹
Celery

西芹有非常顯著的降血壓、降血脂功能，亦含豐富的纖維素，特別適合肥胖人士食用。

Rich in dietary fibre, it can lower blood pressure and blood lipids significantly. It is especially suitable for fat people.

韭黃
Yellow Chives

韭黃易熟，烹調時應最後加入，如果煮得太久釋出水分，可加入適量生粉水打獻。

They should be added last as they would be over-cooked easily. If they are cooked too long and water is released, thickening glaze can be added.

蓮藕
Lotus Root

購買蓮藕時，應選擇體型粗壯、肉質粉嫩結實、已洗去污泥的。

Pick lotus root that is thick in shape, the flesh is firm in texture and the mud is cleaned.

馬鈴薯
Potato

已出芽的馬鈴薯不可食用，否則會出現噁心、腹瀉、發熱等中毒徵狀。

Potato cannot be used if it is sprouted, otherwise you may feel nauseous, have diarrhoea or fever.

栗子
Chestnut

栗子含大量澱粉質，每次進食的分量不宜過多，以免出現脹滯的情況。另外，栗子以外形飽滿的為佳。

It is rich in starch. Don't eat too much, otherwise you may have indigestion. Pick chestnut that is firm and solid.

洋葱
Onion

切洋葱前可以在刀上灑一點冰水，這樣能溶解洋葱刺眼的物質，避免刺激淚腺。

Sprinkle some iced water on the knife before cutting the onion. This can dissolve the eye-irritating substances and avoid irritating the lachrymal glands.

乾冬菇
Dried Black Mushrooms

乾冬菇以外形整齊、色澤自然、帶香濃菇味的為佳。購買後最好密封貯存。

Pick those with regular shape, natural colour and strong mushroom flavour. Store them in air-tight container.

蝦
Shrimp

蝦肉營養豐富，而且脂肪含量較低；但蝦卵和蝦頭有很高膽固醇，不宜多吃。

It is nutritious and low in fat. The eggs and head of a shrimp have high cholesterol level. Avoid eating too much.

魷魚
Squid

魷魚含豐富蛋白質，有補虛損、益氣血的功效，適合身體虛弱人士食用。

It is rich in protein. It can strengthen the body and is suitable for those who are weak.

鹹蛋已有足夠鹹味，所以不須下鹽調味。

No need to season with salt. The salted egg is salty enough.

鹹 蛋 蒸 肉 餅

Steamed Meat Patty with Salted Egg

Ingredients

免治豬肉 5 兩，鹹蛋 1 個

190 g minced pork, 1 salted egg

Seasoning

生粉、油、糖各 1 茶匙，水 3 茶匙

1 tsp caltrop starch, 1 tsp oil, 1 tsp sugar,
3 tsps water

Method

1. 免治豬肉加入調味料和鹹蛋白，以同一方向拌勻。放在碟上，加上鹹蛋黃。
2. 燒滾水，用大火蒸 15 分鐘即成。

1. Mix minced pork with seasoning and salted egg white and stir in one direction. Put them on a plate and place the egg yolk on the top.
2. Bring the water to the boil. Steam over high heat for 15 minutes. Serve.

冬菇馬蹄蒸肉餅
Minced Pork with Mushrooms and Water Chestnuts

Ingredients

免治豬肉 5 兩，冬菇 4 朵，馬蹄 4 粒

190 g minced pork, 4 dried black mushrooms, 4 water chestnuts

Seasoning

生抽、生粉、糖、油各少許

light soy sauce, caltrop starch, sugar, oil

Method

1. 冬菇用水浸軟，擠出水分。
2. 冬菇、糖及油拌勻醃片刻。
3. 馬蹄去皮，和冬菇一同切粒。
4. 免治豬肉加入調味料和適量清水，拌勻，最後加入馬蹄和冬菇。
5. 燒滾水，用猛火蒸 15 分鐘，即可食用。

1. Soak dried black mushrooms in water. Sqeeze out the water.
2. Mix mushrooms with sugar and oil. Set aside.
3. Peel off the skin of water chestnuts. Dice water chestnuts and mushrooms.
4. Mix minced pork with seasoning and add some water. Stir until mixed. Lastly, add water chestnuts and mushrooms.
5. Bring the water to the boil. Steam over high heat for 15 minutes. Serve.

生 炒 排 骨
Sweet and Sour Spareribs

Ingredients

排骨 3 條，青甜椒 1 個，洋葱半個，菠蘿 3 片

3 strips bone-in spareribs, 1 green bell pepper,
1/2 onion, 3 slices pineapple

Marinade

生粉、糖、生抽各 1 茶匙，蛋白 1 個，
白胡椒粉少許

1 tsp caltrop starch, 1 tsp sugar, 1 tsp light soy sauce,
1 egg white, ground white pepper

Sauce

茄汁 3/4 杯，醋半杯，片糖半塊，生粉 1 茶匙

3/4 cup tomato sauce, 1/2 cup vinegar,
1/2 piece raw cane sugar, 1 tsp caltrop starch

Method

1. 排骨斬件，加醃料醃 2-3 小時。
2. 洋葱、青甜椒及菠蘿切塊。
3. 排骨炸至呈金黃色。
4. 洋葱及甜青椒炒熟，盛起。
5. 煮滾獻汁，加入排骨、洋葱、青甜椒及菠蘿，拌勻即成。

1. Cut spareribs into chunks. Add marinade and leave it for 2-3 hours.
2. Cut onion, green bell pepper and pineapple into wedges.
3. Deep fry spareribs until golden colour.
4. Stir fry onion and bell peppers. Set aside.
5. Bring the sauce to the boil. Add spareribs, onion, green bell pepper and pineapple. Serve.

上海排骨
Spareribs in Shanghaiese Style

Ingredients

排骨 1 至 1 1/2 斤

600-900 g spareribs

Marinade

老抽 1 湯匙，糖 3 湯匙，生粉、紹酒各 1 茶匙，蛋白 1 個

1 tbsp dark soy sauce, 3 tbsps sugar, 1 tsp caltrop starch, 1 tsp Shaoxing wine, 1 egg white

Sauce

片糖 1 塊，白醋 1 杯

1 piece raw cane sugar, 1 cup white vinegar

Method

1. 排骨加醃料拌匀，醃 3-4 小時。
2. 排骨炸至熟透，備用。
3. 煮滾獻汁至稠，加入排骨拌匀即成。

1. Add marinade to spareribs and leave it for 3-4 hours.
2. Deep fry spareribs until well done. Set aside.
3. Bring the sauce to the boil until thickened and mix with spareribs. Serve.

這道菜式口味清新，特別適合夏季食用。

The taste of this dish is refreshing. It is especially suitable for summer.

青瓜菠蘿炒排骨

Stir Fried Spareribs with Cucumber and Pineapple

Ingredients

排骨 1 斤，青瓜半條，菠蘿 3 片，紅甜椒半個

600 g spareribs, 1/2 cucumber, 3 slices pineapple, 1/2 red bell pepper

Marinade

生抽、糖、生粉各 1 茶匙，麻油少許，蛋白 1 個，胡椒粉少許

1 tsp light soy sauce, 1 tsp sugar, 1 tsp caltrop starch, sesame oil, 1 egg white, ground white pepper

Sauce

白醋半杯，生抽半湯匙，糖 1 茶匙，蒜頭 2 粒(剁茸)

1/2 cup white vinegar, 1/2 tbsp light soy sauce, 1 tsp sugar, 2 cloves garlic (finely chopped)

Method

1. 排骨加醃料拌勻，備用。
2. 青瓜洗淨後切塊；菠蘿片切塊；紅椒去籽及切塊。
3. 排骨炸至呈金黃色及熟透，瀝乾油分。
4. 煮滾獻汁，加入排骨、青瓜、紅椒和菠蘿拌炒片刻即成。

1. Add marinade to spareribs. Set aside.
2. Rinse cucumber and cut into wedges. Cut pineapple into chunks. Seed red pepper and cut into wedges.
3. Deep fry spareribs until golden brown and done. Drain.
4. Bring the sauce to the boil. Add spareribs, cucumber, red pepper and pineapple. Stir fry. Serve.

蒜 茸 豆 豉 蒸 排 骨
Steamed Spareribs with Fermented Black Beans and Garlic

Ingredients

排骨 1 斤，豆豉 2 湯匙，蒜頭 5 粒

600 g spareribs, 2 tbsps fermented black beans,
5 cloves garlic

Marinade

生抽 1 茶匙，糖、生粉各半茶匙

1 tsp light soy sauce, 1/2 tsp sugar,
1/2 tsp caltrop starch

Method

1. 排骨洗淨，斬件。
2. 蒜頭剁成茸；豆豉略椿爛。
3. 排骨、蒜茸、豆豉和醃料拌勻。
4. 把所有材料放在碟上，水滾後蒸約 15-20 分鐘即成。

1. Rinse spareribs. Chop into chunks.
2. Grate garlic finely. Pound black beans lightly.
3. Mix spareribs with garlic, black beans and marinade. Set aside.
4. Put all ingredients on a plate. Bring the water to the boil. Steam for about 15-20 minutes. Serve.

酥 炸 豬 扒
Deep Fried Pork Chops

Ingredients

豬扒 4 塊，麵包糠 1 杯，蛋液適量

4 pieces pork chop, 1 cup breadcrumbs, whisked egg

Marinade

生抽、糖各半茶匙，生粉 1 茶匙，水 3 湯匙，麻油、胡椒粉、蛋白各少許

1/2 tsp light soy sauce, 1/2 tsp sugar, 1tsp caltrop starch, 3 tbsps water, sesame oil, ground white pepper, egg white

Method

1. 豬扒洗淨，用刀背拍鬆，加醃料拌勻，醃 2-3 小時。
2. 豬扒蘸上已拂勻的蛋液，再均勻地沾上麵包糠。
3. 用慢火炸熟及兩面呈金黃色，盛起，瀝乾油分，切塊後可蘸沙律醬食用。

1. Rinse pork chops and pound with the back of a knife until soft. Mix with marinade and leave for 2-3 hours.
2. Mix pork chops with whisked egg. Coat with breadcrumbs evenly.
3. Deep fry over low heat until both sides are golden and done. Drain and cut into pieces. Serve with salad dressing.

煮熟的洋葱帶甜味，小朋友同樣喜愛。

Cooked onions are sweet. Children would like it much.

洋 葱 豬 扒

Fried Pork Chops with Onion

Ingredients

豬扒 4 塊，洋葱 1 個

4 pieces pork chop,1 onion

Marinade

生抽、糖、生粉各 1 茶匙，麻油、胡椒粉各少許

1 tsp light soy sauce, 1 tsp sugar,
1 tsp caltrop starch, sesame oil, ground white pepper

Sauce

蠔油 3 茶匙，糖半茶匙，水半杯

3 tsps oyster sauce, 1/2 tsp sugar, 1/2 cup water

Method

1. 洋葱洗淨，切條。
2. 豬扒洗淨，用刀背拍鬆，加醃料拌勻，醃 2-3 小時。燒熱油，下豬扒煎至熟透。
3. 燒熱油，加入洋葱略炒，加入豬扒拌勻。
4. 注入獻汁，略煮後即可食用。

1. Rinse and cut onion into strips.
2. Rinse pork chops and pound with the back of a knife until soft. Mix with marinade and leave for 2-3 hours. Heat oil, fry pork chops until done.
3. Heat oil and stir fry onion for a while, mix with pork chops.
4. Pour the sauce, cook for a while. Serve.

咖 喱 豬 扒
Curry Pork Chops

豬扒 4 塊

4 pieces pork chop

咖喱粉 1 包，雞蛋 1 隻(拂勻)，薑 2 片，蔥 2 條，生粉 1 湯匙，生抽 1 茶匙，紹酒半茶匙

1 pack curry powder, 1 whisked egg, 2 slices ginger, 2 sprigs spring onion, 1 tbsp caltrop starch, 1 tsp light soy sauce, 1/2 tsp Shaoxing wine

1. 豬扒洗淨，用刀背拍鬆，加醃料拌勻，醃 2-3 小時。
2. 燒熱大量油，用慢火把豬扒炸熟及兩面呈金黃色，盛起，瀝乾油分，切塊後即可食用。

1. Rinse pork chops and pound with the back of a knife until soft. Mix with marinade and leave for 2-3 hours.
2. Heat a large amount of oil, deep fry pork chops over low heat until both sides are golden and done. Drain and cut into pieces. Serve.

茄汁菠蘿豬扒
Pork Chops with Pineapple in Tomato Sauce

Ingredients

豬扒 4 塊，菠蘿片 1 小罐，甜紅椒 1 個

4 pieces pork chop, 1 small can pineapple slices,
1 red bell pepper

Marinade

雞蛋 1 隻，薑 2 片，生抽 1 茶匙，
糖半茶匙，紹酒半湯匙，生粉 1 湯匙，
麻油、胡椒粉各少許

1 egg, 2 slices ginger, 1 tsp light soy sauce, 1/2 tsp sugar, 1/2 tbsp Shaoxing wine, 1 tbsp caltrop starch, sesame oil, ground white pepper

Sauce

茄汁 1 杯，水半杯，糖 2 茶匙，生抽 1 茶匙

1 cup tomato sauce, 1/2 cup water, 2 tsps sugar, 1 tsp light soy sauce

Method

1. 豬扒洗淨，用刀背拍鬆，加醃料拌勻，醃 3-4 小時。
2. 把 1 片菠蘿切成 3 塊。紅椒切成相同大小。
3. 燒熱油，把豬扒炸熟及兩面呈金黃色，盛起，瀝乾油分。
4. 加熱獻汁，煮至濃稠，加入所有材料，略炒即可食用。

1. Rinse pork chops and pound with the back of a knife until soft. Mix with marinade, leave for 3-4 hours.
2. Cut pineapple slices into 3 pieces. Cut the bell pepper as the size of pineapple.
3. Heat oil and deep fry pork chops until golden and done. Drain.
4. Bring the sauce to the boil and cook until thickened. Mix with all ingredients, stir fry for a while. Serve.

南乳蓮藕燜腩肉
Stewed Pork Belly with Lotus Root and Fermented Tarocurd

Ingredients

半肥瘦豬肉 1 斤，蓮藕 1 個，南乳 1 塊，蒜頭 2 粒(切碎)，冰糖 1 茶匙(切碎)

600 g half fatty pork belly, 1 lotus root, 1 piece fermented tarocurd, 2 cloves garlic (chopped), 1 tsp rock sugar (chopped)

Method

1. 豬肉洗淨，切塊。
2. 蓮藕洗淨，去皮，切片。
3. 燒熱油，加入蒜頭及南乳略炒，下豬肉及蓮藕炒一會。加入適量清水煮滾，煮至豬肉熟透。
4. 加入冰糖，轉中火煮至蓮藕腍身，即可食用。

1. Rinse and cut pork belly into pieces.
2. Rinse lotus root, peel off the skin and slice.
3. Heat oil and stir fry garlic and fermented tarocurd. Add lotus root and stir fry for a while. Pour some water. Bring to the boil again. Cook until pork belly done.
4. Add rock sugar and turn to medium heat. Cook lotus root until tender. Serve.

如果擔心這道菜太肥膩，可按喜好減少燒腩的分量。

If you think this dish is too fatty, the quantity of pork belly can be adjusted.

冬 瓜 枝 竹 燜 火 腩

Stewed Roast Pork with Winter Melon and Deep Fried Beancurd Stick

Ingredients

燒腩 6 兩(切小塊)，冬瓜半斤，枝竹 1 條，蒜頭 2 粒

225 g roast pork (chopped), 300 g winter melon, 1 piece deep fried beancurd stick, 2 cloves garlic

Seasoning

蠔油、生抽各 1 茶匙，水 3 茶匙

1 tsp oyster sauce, 1 tsp light soy sauce, 3 tsps water

Method

1. 冬瓜去皮，洗淨，切成和燒腩一樣大小。
2. 枝竹浸軟，切段。
3. 燒熱油，炒香蒜頭和燒腩，加入冬瓜略炒。
4. 加入適量清水、調味料及枝竹，燜 10 分鐘即成。

1. Peel winter melon. Rinse and cut into pieces with the same size of roast pork.
2. Soak beancurd stick in the water until soft. Cut into sections.
3. Heat oil, stir fry garlic and roast pork until fragrant. Add winter melon, stir fry for a while.
4. Add some water, seasoning and beancurd stick. Simmer for about 10 minutes. Serve.

馬鈴薯燜腩肉
Stewed Pork Belly with Potatoes

豬腩肉 6 兩，馬鈴薯 2 個，乾葱頭 1 粒，
蒜頭 1 粒，香葉 1 片，紹酒 1 茶匙

225 g fatty pork belly, 2 potatoes, 1 clove shallot,
1 clove garlic, 1 piece bay leaf, 1 tsp Shaoxing wine

生抽及老抽各少許，黑胡椒 1 茶匙，
冰糖 1 湯匙(切碎)

light soy sauce, dark soy sauce, 1 tsp black peppercorns,
1 tbsp rock sugar (chopped)

1. 豬腩肉洗淨，切塊。
2. 馬鈴薯去皮，洗淨，切塊。
3. 燒熱油，馬鈴薯略炸，盛起，瀝乾油分待用。
4. 用少許油爆香乾葱頭和蒜頭，加入豬腩肉拌勻，灒酒，再拌勻，加入調味料、香葉和馬鈴薯略拌。
5. 加入適量清水，用猛火煮滾，轉慢火燜 45 分鐘即成。

1. Rinse and cut pork belly into chunks.
2. Peel, rinse potatoes. Cut into pieces.
3. Heat oil, deep fry potatoes for a while. Drain and set aside.
4. Leave a little oil, put garlic and shallot till fragrant. Add pork belly. Stir fry and sizzle with wine. Stir fry again. Add seasoning, bay leaf and potatoes.
5. Pour the water. Bring to the boil. Turn to low heat and simmer for 45 minutes. Serve.

燜豬肉加入冰糖，令肉質較軟和縮短烹調時間。

Rock sugar is added, the pork will be very soft and the cooking time will be shortened.

梅菜扣肉
Stewed Pork Belly with Mui Choi

Ingredients

豬腩肉半斤，梅菜 1 棵，蒜頭 2 粒，
紹酒 1 湯匙

300 g pork belly, 1 Mui Choi (preserved flowering cabbage), 2 cloves garlic, 1 tbsp Shaoxing wine

Seasoning

生抽、老抽、冰糖各適量

light soy sauce, dark soy sauce, rock sugar

Method

1. 豬腩肉洗淨，切塊。
2. 梅菜洗淨，用清水略浸，切小片。
3. 燒熱油，加入蒜頭爆香，下豬腩肉略炒，灒酒，炒勻，加入調味料和梅菜炒勻。
4. 加入過面清水，用猛火煮滾，轉慢火燜 45-60 分鐘，即可食用。

1. Rinse and cut pork belly into chunks.
2. Rinse and soak Mui Choi in water for a while. Cut into small pieces.
3. Heat oil, put garlic and stir fry until fragrant. Add pork belly, stir fry for a while, sizzle with wine and stir fry again. Add seasoning and Mui Choi, stir fry briefly.
4. Pour water to cover the ingredients. Bring to the boil, turn to low heat and simmer for 45 minutes to 1 hour. Serve.

蔬菜雜粒炒肉丁
Stir Fried Diced Meat and Vegetables

豬瘦肉 3 兩，乾冬菇 4 朵，粟米芯 4 條，
紅蘿蔔粒 1 杯，豆腐 1 塊

113 g lean pork, 4 dried mushrooms, 4 pieces baby corn,
1 cup diced carrot, 1 piece beancurd

生抽、糖、生粉各 1 茶匙，胡椒粉、麻油各少許

1 tsp light soy sauce, 1 tsp sugar, 1 tsp caltrop starch,
ground white pepper, sesame oil

鹽、麻油各適量

salt, sesame oil

1. 粟米芯及紅蘿蔔洗淨，切粒。
2. 豬肉和醃料拌勻，醃 2-3 小時。
3. 冬菇浸軟，去蒂，瀝乾水分，切粒。
4. 豆腐吸乾水分，灑少許鹽，用油炸至金黃色，瀝乾油分，切粒。
5. 燒熱油，下豬肉炒至九成熟，加入其他材料炒片刻，加入調味料略炒，即可上碟食用。

1. Rinse and dice baby corn and carrot.
2. Mix lean pork with marinade. Leave for 2-3 hours.
3. Soak dried mushrooms until soft. Remove the stalks. Drain and cut into cubes.
4. Wipe dry the beancurd. Sprinkle with salt on the beancurd. Deep fry beancurd until golden. Drain and dice.
5. Heat oil. Stir fry lean pork until almost done. Add remaining ingredients and stir fry for a while. Add seasoning and stir well. Serve.

蘿蔔肉片炒蒜心
Stir Fried Sliced Meat with Radish and Garlic Stems

Ingredients

豬瘦肉 4 兩，白蘿蔔(小) 1 條，蒜心 1 束

150 g lean pork, 1 radish (small size),
1 bundle garlic stems

Marinade

生抽、糖各半茶匙，生粉 1 茶匙，胡椒粉少許

1/2 tsp light soy sauce, 1/2 tsp sugar,
1 tsp caltrop starch, ground white pepper

Method

1. 白蘿蔔去皮，切條。
2. 蒜心洗淨，切段。
3. 豬肉洗淨略醃，下油略炒，盛起。
4. 燒熱油，下白蘿蔔和蒜心炒至半熟，加入豬肉炒至熱透，灑少許鹽調味，即可食用。

1. Peel off the skin of radish, cut into strips.
2. Rinse garlic stems and cut into sections.
3. Rinse and marinate lean pork. Heat oil, stir fry for a while and set aside.
4. Heat the oil. Stir fry radish and garlic stems until half-cooked. Add lean pork and stir fry until done. Sprinkle with salt and serve.

番茄馬蹄肉丸
Meat Balls with Tomato and Water Chestnut

免治豬肉半斤，番茄 1 個，馬蹄 6 粒，黃甜椒半個

300 g minced pork, 1 tomato, 6 water chestnuts,
1/2 yellow bell pepper

生抽 1 茶匙，生粉半茶匙，麻油少許

1 tsp light soy sauce, 1/2 tsp caltrop starch, sesame oil

茄汁 4 湯匙，生抽半茶匙，水半杯

4 tbsps tomato sauce, 1/2 tsp light soy sauce,
1/2 cup water

1. 番茄洗淨，切塊；黃椒洗淨，去籽、切塊。
2. 免治豬肉和醃料拌勻，醃 2-3 小時。
3. 馬蹄去皮，切碎，加入免治豬肉拌勻。
4. 把豬肉搓成肉丸，下油炸至呈金黃色。
5. 燒熱油，加入番茄和獻汁拌勻，加入肉丸煮熟，最後下黃椒拌炒即可食用。

1. Rinse and cut tomato into pieces. Rinse yellow bell pepper. Seed and cut into pieces.
2. Mix minced pork with marinade. Leave for 2-3 hours.
3. Peel off the skin of water chestnuts, finely chop and mix with minced pork.
4. Knead minced pork into balls. Heat oil and deep fry until golden.
5. Heat oil, stir fry tomato and the sauce, mix well and add the meat balls and yellow pepper. Stir fry until done. Serve.

如果使用咖喱粉，要先和椰漿徹底拌勻，否則容易結成粉粒。

Mix curry powder with coconut milk well before use. Otherwise, it gets lumpy.

咖喱牛腩
Curry Beef Flank

Ingredients

牛腩 1 斤，馬鈴薯 2 個，洋葱 1 個，紅蘿蔔 1 條，紅甜椒、青甜椒各 1 個，咖喱粉/咖喱醬 2 茶匙，椰漿半杯

600 g beef flank, 2 potatoes, 1 onion, 1 carrot, 1 red bell pepper, 1 green bell pepper, 2 tsps curry powder or curry paste, 1/2 cup coconut milk

Seasoning

鹽 1 茶匙

1 tsp salt

Method

1. 馬鈴薯和紅蘿蔔去皮，切塊。洋葱切塊；青、紅椒洗淨，去籽、切塊。
2. 牛腩飛水，用水沖淨。
3. 燒熱鑊，用油 2 湯匙爆香洋葱，加入牛腩、紅蘿蔔及馬鈴薯炒香。
4. 加入咖喱粉和椰漿煮滾，轉慢火燜至九成熟，最後加入紅椒及青椒，灑入鹽煮至全熟即成。

1. Peel off the skin of potatoes and carrot. Rinse onion, red and green pepper. Seed the peppers. Cut all of them into pieces.
2. Scald beef in boiling water and rinse well.
3. Heat a wok and add 2 tbsps oil. Add onion and stir fry until fragrant. Add beef, carrot and potatoes and stir fry for a while.
4. Add curry powder and coconut milk. Bring to the boil. Simmer until almost done. Add red and green bell pepper and cook for a while. Sprinkle with salt. Serve.

榨菜蒸牛肉
Steamed Beef with Zha Cai

牛肉 5 兩，榨菜 1 個

190 g beef, 1 preserved Sichuan vegetable (Zha Cai)

生抽、生粉各 1 茶匙，油少許

1tsp light soy sauce, 1 tsp caltrop starch, oil

1. 榨菜洗淨，切薄片，用水泡 10 分鐘，擠乾水分。
2. 牛肉切薄片，用醃料拌勻，略醃。
3. 把牛肉和榨菜放在碟上，水滾後，蒸約 10 分鐘，即可食用。

1. Rinse Zha Cai and slice thinly. Soak it in the water for 10 minutes. Squeeze the water.
2. Slice beef thinly. Mix beef with marinade, leave for a while.
3. Arrange the beef and Zha Cai on a plate. Bring the water to the boil. Steam for about 10 minutes. Serve.

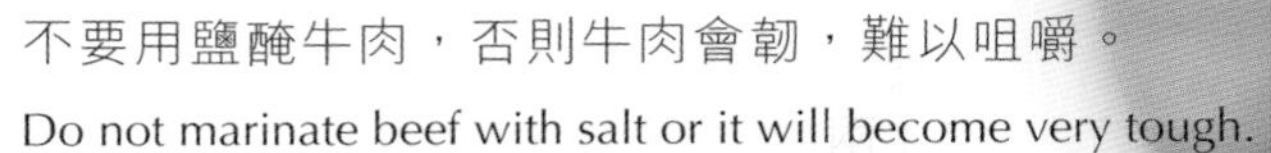

不要用鹽醃牛肉，否則牛肉會韌，難以咀嚼。

Do not marinate beef with salt or it will become very tough.

三色甜椒炒牛肉

Stir Fried Beef with Bell Peppers

紅甜椒、青甜椒、黃甜椒各 1/4 個，牛肉 4 兩

1/4 red bell pepper, 1/4 green bell pepper, 1/4 yellow bell pepper, 150 g beef

生抽半湯匙，油、生粉各半茶匙，麻油、胡椒粉各少許

1/2 tbsp light soy sauce, 1/2 tsp oil, 1/2 tsp caltrop starch, sesame oil, ground white pepper

麻油、鹽適量

sesame oil, salt

1. 牛肉洗淨，切片，加入醃料拌勻，醃 2-3 小時。
2. 甜椒洗淨，切條。燒熱油，炒熟甜椒，盛起。
3. 燒熱油，下牛肉炒熟，加入甜椒及調味料再炒勻即成。

1. Rinse beef and slice. Mix with marinade. Set aside for 2-3 hours.
2. Rinse bell peppers and shred. Heat wok with oil. Stir fry peppers until done. Set aside.
3. Heat oil, stir fry beef until done. Add bell peppers and seasoning. Stir well. Serve.

栗子燜雞
Stewed Chicken with Chestnuts

Ingredients

雞 1 隻，栗子半斤，蒜頭 2 粒，薑 1 片，紹酒 1 茶匙

1 whole chicken, 300 g chestnuts, 2 cloves garlic, 1 slice ginger, 1 tsp Shaoxing wine

Seasoning

生抽、冰糖碎各 1 茶匙，老抽 2 茶匙

1 tsp light soy sauce, 1 tsp rock sugar (chopped), 2 tsps dark soy sauce

Method

1. 雞洗淨，抹乾水分，斬件。
2. 栗子去殼、去衣。
3. 燒熱 1 湯匙油，爆香薑片和蒜頭，加入雞塊略炒，灒酒，加入栗子拌勻。
4. 加入調味料，注入水（蓋過材料），用猛火煮滾後再用中火燜 45 分鐘，即可食用。

1. Wash chicken and wipe dry. Chop up.
2. Remove shell and skin of chestnuts.
3. Heat 1 tbsp of oil, stir fry ginger and garlic until fragrant. Add chicken, stir fry for a while. Sizzle wine. Add chestnuts. Mix well.
4. Add seasoning and water (water level should be sufficient to cover all ingredients). Bring to the boil over high heat. Then turn to medium heat and simmer for 45 minutes. Serve.

南瓜燜雞
Stewed Chicken with Pumpkin

Ingredients

雞半隻，南瓜半個，蒜頭 2 粒，水 1 杯

1/2 chicken, 1/2 pumpkin, 2 cloves garlic, 1 cup water

Marinade

生抽 1 茶匙，生粉半茶匙，胡椒粉少許

1 tsp light soy sauce, 1/2 tsp caltrop starch, ground white pepper

Method

1. 雞洗淨後抹乾，切大塊，下醃料拌勻待用。
2. 南瓜去皮，切件。
3. 燒熱油，爆香蒜頭，加入雞塊略炒，加入南瓜和水。
4. 用中火燜 15 分鐘，下鹽調味即可食用。

1. Rinse chicken and wipe dry. Chop up. Mix with marinade. Set aside for later use.
2. Peel off the skin of pumpkin and cut into pieces.
3. Heat oil, stir fry garlic until fragrant. Put in chicken and stir fry for a while. Add pumpkin and water.
4. Simmer over medium heat for 15 minutes. Season with salt. Ready to serve.

腰果西芹雞柳

Stir Fried Chicken Fillet with Celery and Cashew Nuts

Ingredients

雞柳半斤，西芹 3 枝，腰果半杯，紅蘿蔔半條，蒜頭 1 粒

300 g chicken fillet, 3 stalks celery, 1/2 cup cashew nuts, 1/2 carrot, 1 clove garlic

Marinade

生抽 1 茶匙，糖、生粉各半茶匙，油、麻油各 1/4 茶匙，
胡椒粉、蒜頭 1 粒 (剁茸)

1 tsp light soy sauce, 1/2 tsp sugar, 1/2 tsp caltrop starch, 1/4 tsp oil, 1/4 tsp sesame oil, ground white pepper, 1 clove garlic (crushed)

Seasoning

鹽少許

salt

Method

1. 煮滾清水，加入 1/4 茶匙鹽和腰果，煮 1 分鐘，盛起腰果，用廚房紙抹乾。
2. 把油煮暖，加入腰果，用中火炸至呈淺啡色，盛起，瀝乾油分。
3. 雞柳洗淨後抹乾，與紅蘿蔔、西芹分別切成幼條；用醃料醃雞柳。
4. 燒熱油，下紅蘿蔔和西芹拌炒，盛起待用。
5. 燒熱鑊，下 2 湯匙油爆香蒜頭，下雞柳炒至熟透，加入紅蘿蔔和西芹炒勻，下鹽調味，加入腰果拌勻即可享用。

1. Bring water to the boil. Add 1/4 tsp of salt and cashew nuts. Cook for 1 minute. Remove and drain. Wipe dry with kitchen paper.
2. Put cashew nuts in warm oil. Deep fry over medium heat until light brown. Remove and drain oil.
3. Rinse chicken fillet and wipe dry. Cut chicken fillet, carrot and celery into shreds separately. Mix chicken fillet with marinade.
4. Heat oil. Put in carrot and celery. Stir fry until done. Remove and set aside.
5. Heat wok and add 2 tbsps of oil. Stir fry garlic until fragrant. Add chicken fillet. Stir fry until done. Put in carrot and celery. Stir well. Add salt to taste. Put in cashew nuts. Stir well and serve.

酥炸雞翼
Deep Fried Chicken Wings

Ingredients

雞翼 8 隻，雞蛋 1 個 (拂勻)，麵包糠適量

8 chicken wings, 1 egg (whisked), breadcrumbs

Marinade

糖、生抽、生粉各 1 茶匙，胡椒粉少許

1 tsp sugar, 1 tsp light soy sauce, 1 tsp caltrop starch, ground white pepper

Method

1. 雞翼洗淨，抹乾水分，加入醃料醃約 30 分鐘。雞蛋與雞翼拌勻，均勻地沾上麵包糠。
2. 燒熱大量油，把雞翼炸熟及呈金黃色即成。

1. Rinse chicken wings and wipe dry. Mix with marinade and marinate for 30 minutes. Coat chicken wings with whisked egg and then breadcrumbs evenly.
2. Heat large amount of oil. Deep fry chicken wings until golden and done. Serve.

將已浸冬菇用糖和油醃勻，只比用清水浸透，味道會更佳。

The taste of dried black mushrooms is better by marinading in sugar and oil. This would be better by soaking them in water only.

冬菇雲耳蒸雞中翼

Steamed Chicken Wings with Cloud Ears Fungus and Mushrooms

Ingredients

雞中翼 6 隻，冬菇 3 朵，雲耳 10 朵，
糖、油各少許

6 mid-joint chicken wings, 3 dried black mushrooms, 10 dried cloud ears, sugar, oil

Marinade

生抽、生粉各 1 茶匙，麻油、糖各 1/4 茶匙

1 tsp light soy sauce, 1 tsp caltrop starch, 1/4 tsp sesame oil, 1/4 tsp sugar

Method

1. 冬菇浸透後，擠乾水分，切絲，再用糖和油醃 4 小時。
2. 雲耳浸透，洗淨，瀝乾水分；雞翼切件。
3. 雞中翼用醃料醃勻，加入冬菇和雲耳，水滾後蒸 12-15 分鐘即成。

1. Soak dried black mushrooms in water until soft. Squeeze out the water. Shred. Marinate mushrooms with sugar and oil for 4 hours.
2. Soak cloud ears until soft. Rinse and drain. Chop up chicken wings.
3. Mix chicken wings with marinade. Add mushrooms and cloud ears. Bring the water to the boil. Steam for 12-15 minutes. Serve.

檸 檬 雞 槌
Lemon Chicken Mallets

雞中翼 8 隻

8 mid-joint chicken wings

檸檬(榨汁)、蛋白各 1 個，生粉 1 茶匙，生抽、糖各半茶匙

juice of 1 lemon, 1 egg white, 1 tsp caltrop starch, 1/2 tsp light soy sauce, 1/2 tsp sugar

1. 雞中翼洗淨，除去其中一塊骨，把肉捲向另一邊，做成雞腿形狀。
2. 雞翼加入醃料拌勻，醃 2-3 小時。
3. 用平底鍋燒熱油，把雞翼炸至呈金黃色，盛起即可食用。

1. Rinse chicken wings. Remove one of the bones. Fold the flesh down to sharp like a mallet.
2. Mix chicken wings with marinade. Leave for 2-3 hours.
3. Heat oil in a frying pan. Deep fry chicken wings until golden brown. Serve.

可用雞湯代替水。

You can use chicken broth instead of water.

蒸滑蛋
Steamed Egg Custard

雞蛋 3 個，水 2 杯，鹽適量

3 eggs, 2 cups water, salt

1. 雞蛋拂勻，加入水和鹽，拌勻。
2. 把蛋液注入深碟中，蓋上錫紙。
3. 燒滾水，用大火蒸約 8 分鐘即成。

1. Whisk eggs with water, mix well. Add salt.
2. Pour the egg mixture into a deep plate. Cover with aluminum foil.
3. Bring the water to the boil. Steam over high heat for about 8 minutes. Serve.

番茄炒蛋
Stir Fried Tomatoes and Eggs

Ingredients

番茄 2 個，雞蛋 3 個，茄汁 5 湯匙，糖、鹽各少許

2 tomatoes, 3 eggs, 5 tbsps tomato sauce, sugar, salt

Method

1. 番茄洗淨，每個番茄切成 4-6 塊。雞蛋拂勻。
2. 熱鑊下油，將雞蛋炒至半熟，備用。
3. 熱鑊下油，將番茄炒至軟身，加入炒蛋和茄汁。
4. 加入糖和鹽，拌勻即成。

1. Rinse tomatoes. Cut each tomato into 4-6 pieces. Whisk eggs.
2. Heat wok with some oil. Stir fry the eggs until half done. Set aside.
3. Heat wok with some oil. Stir fry tomatoes until tender. Mix with scrambled eggs and add tomato sauce.
4. Add sugar and salt to taste and stir well. Serve.

煎薯仔蛋餅
Fried Shredded Potato and Eggs

Ingredients

馬鈴薯 1 個，雞蛋 5 個

1 potato, 5 eggs

Seasoning

鹽適量

salt

Method

1. 雞蛋拂勻。
2. 馬鈴薯去皮，切小塊。
3. 燒熱油，把馬鈴薯炒至九成熟，盛起備用。
4. 馬鈴薯和雞蛋拌勻，加入調味料，拌勻。
5. 燒熱油，注入蛋液及馬鈴薯，煎至淺啡色，翻轉另一邊，煎好後即可食用。

1. Whisk eggs well.
2. Peel off the skin of potato. Cut into small pieces.
3. Heat oil. Stir fry potato until almost done. Remove and set aside.
4. Mix potato and whisked eggs. Add seasoning and mix well.
5. Heat oil. Pour the mixture of eggs and potato, fry until lightly brown. Turn over and fry again. Serve.

叉燒洋蔥炒蛋
Stir Fried BBQ Pork with Egg and Onion

Ingredients

雞蛋 5 個，洋蔥半個，叉燒半斤

5 eggs, 1/2 onion, 300 g BBQ pork

Method

1. 洋蔥去衣，切小片。雞蛋拂勻。
2. 燒熱 1 茶匙油，爆香洋蔥。
3. 加入叉燒和已拂勻的雞蛋，炒勻，炒至熟透即可。

1. Peel off the skin of onion and cut into small pieces. Whisk egg.
2. Heat 1 tsp of oil and stir fry onion until fragrant.
3. Add BBQ pork and whisked egg. Stir fry until done. Serve.

粟 米 龍 脷 柳
Sole Fillet in Corn Sauce

Ingredients

龍脷柳 6 兩，忌廉粟米 1 罐，
雞蛋 1 個(拂勻)，水半杯

225 g fish fillet, 1 can cream style sweet corn, 1 whisked egg, 1/2 cup water

Marinade

薑茸 2 湯匙，紹酒 1 茶匙，胡椒粉少許，
生粉 2 茶匙（後下）

2 tbsps chopped ginger, 1 tsp Shaoxing wine, ground white pepper, 2 tsps caltrop starch (added lastly)

Method

1. 魚柳洗淨，切件，加入醃料拌勻略醃，然後撲上生粉。
2. 燒熱油，把魚柳炸至金黃色，盛起，瀝乾油分，置於碟上。
3. 煮滾忌廉粟米和水，加入已拂勻的雞蛋，一邊加入一邊拌勻，煮熟後澆於魚柳面，即可食用。

1. Rinse and cut fish fillet into pieces. Mix with marinade and set aside for later use. Coat with caltrop starch.
2. Heat oil. Deep fry fish fillet until done. Drain. Arrange on a plate.
3. Bring the sweet corn and water to the boil. Add whisked egg while stirring. Pour onto the fish. Serve.

魚 肉 春 卷
Fish Spring Rolls

Ingredients

鯪魚膠半斤，紅蘿蔔 1 條，葱 1 條，
春卷皮 1 包

300 g minced dace, 1 carrot, 1 sprig spring onion,
1 pack spring roll wrapper

Seasoning

生粉 1 茶匙，生抽半茶匙，胡椒粉少許

1 tsp caltrop starch, 1/2 tsp light soy sauce,
ground white pepper

Method

1. 紅蘿蔔去皮，和葱一同切幼絲。
2. 燒熱油，炒勻紅蘿蔔、葱和調味料，炒熟後盛起。待涼後，與魚膠拌勻成餡料。
3. 把餡料鋪在春卷皮上，捲成春卷。
4. 燒熱油，把春卷炸脆及呈金黃色，盛起，瀝乾油分，切件後即可食用。

1. Peel carrot. Cut carrot and spring onion into fine shreds.
2. Heat oil. Stir fry carrot and spring onion until done. Add seasoning. Remove and set aside until cool. Mix with minced dace.
3. Put the mixture on the spring roll wrapper and roll.
4. Heat oil. Deep fry the fish rolls until golden and crispy. Drain and cut into pieces. Serve.

蘿 蔔 炒 魚 餅
Stir Fried Radish with Shredded Fish Patty

Ingredients

白蘿蔔(中) 1 條，鯪魚膠 6 兩

1 medium-size radish, 225 g minced dace

Seasoning

蠔油 1 湯匙，生抽 1 茶匙，水半杯

1 tbsp oyster sauce, 1 tsp light soy sauce, 1/2 cup water

Method

1. 白蘿蔔去皮，切條。
2. 燒熱油，將鯪魚膠煎熟成魚餅，切條備用。
3. 燒熱油，把白蘿蔔炒至九成熟，下調味料，加入魚餅，炒熟即可。

1. Peel off the skin of radish. Cut into shreds.
2. Heat oil. Fry minced dace into patty. Slice and set aside.
3. Heat oil, stir fry radish until almost done. Add seasoning. Pour in sliced fish patty and cook for a while. Serve.

紅蘿蔔芽菜炒魚片

Stir Fried Bean Sprout with Fish Meat and Carrot

Ingredients

芽菜 5 兩，炸魚片 3 兩，紅蘿蔔(中)半條

190 g green bean sprouts,
113 g deep fried fish meat roll,
1/2 medium-size carrot

Seasoning

鹽、糖各少許

salt, sugar

Method

1. 紅蘿蔔、魚片切條。
2. 芽菜洗淨，瀝乾水分。
3. 燒熱油，炒香魚片，加入紅蘿蔔和芽菜，炒匀，加入調味料拌匀，即可食用。

1. Cut carrot and fish meat into strips.
2. Rinse and drain green bean sprouts.
3. Heat oil. Stir fry fish meat till fragrant. Add carrot and bean sprouts. Stir fry for a while. Add seasoning and stir fry. Dish up and serve.

沙葛炒魷魚
Stir Fried Yam Bean with Squid

Ingredients

魷魚 1 隻，沙葛 5 兩，韭菜花一束，薑 3 片

1 squid, 190 g yam bean,
1 bundle flowering Chinese chives,
3 slices ginger

Seasoning

糖半茶匙，鹽、麻油各適量

1/2 tsp sugar, salt, sesame oil

Method

1. 沙葛洗淨後去皮，切條。韭菜花洗淨，切段。
2. 魷魚劏好，洗淨墨汁。撕去外皮，洗淨。在表面切十字，切塊。煲滾水，下 1 片薑，放入魷魚飛水。
3. 燒熱油，加入薑片、沙葛、韭菜花，炒至沙葛九成熟。
4. 加入魷魚炒至熟，加入調味料，拌勻即成。

1. Rinse and peel yam bean. Cut into shreds. Rinse flowering Chinese chives and cut into sections.
2. Gut squid and remove the ink. Tear off the outer skin. Rinse and mark a cross on the skin. Cut into pieces. Boil water with 1 slice of ginger. Scald squid.
3. Heat oil, add gingers slices, flowering Chinese chives and yam bean. Stir fry yam bean until almost done.
4. Add squid, stir fry until done. Add seasoning. Stir well. Serve.

韭黃蛋絲炒蝦球
Stir Fried Shrimp with Egg and Yellow Chives

Ingredients

蝦 4 兩，雞蛋 5 個，韭菜 3 両(切段)，
生粉、鹽各少許

150 g shrimps, 5 eggs, 113 g yellow Chinese chives (sectioned), caltrop starch, salt

Marinade

生粉、胡椒粉各適量，蛋白 1 個

caltrop starch, ground white pepper, 1 egg white

Method

1. 蝦去殼，洗淨，瀝乾，用生粉和鹽抓洗，再次沖淨，瀝乾，用廚房紙抹乾，加入醃料醃一會。
2. 燒熱油，炒熟蝦仁，盛起待用。
3. 燒熱油，把韭黃炒香，加入蝦仁。
4. 雞蛋拂勻，慢慢倒入鍋中，炒至蛋熟，趁熱享用。

1. Shell, rinse and drain shrimps. Rub with caltrop starch and salt. Rinse, drain and wipe dry with kitchen paper. Mix with marinade and leave it for a while.
2. Heat oil. Stir fry shrimps until done. Set aside.
3. Heat oil. Stir fry yellow Chinese chives until fragrant. Add shrimps.
4. Pour in the beaten egg slowly and stir fry until set. Serve hot.

自煮飯堂 靚餸

作者 Author
蔡美娜

策劃/編輯 Project Editor
Forms Kitchen 編輯委員會 Editorial Committee, Forms Kitchen

攝影 Photographer
細權

美術統籌及設計 Art Direction & Design
Me

出版者 Publisher
Forms Kitchen
香港鰂魚涌英皇道1065號 東達中心1305室
Room 1305, Eastern Centre, 1065 King's Road, Quarry Bay, Hong Kong
電話 Tel 2564 7511
傳真 Fax 2565 5539
電郵 Email info@wanlibk.com
網址 Web Site http//www.formspub.com
http//www.facebook.com/formspub

瀏覽網站

會員申請

發行者 Distributor
香港聯合書刊物流有限公司
SUP Publishing Logistics (HK) Ltd.
香港新界大埔汀麗路36號 中華商務印刷大厦3字樓
3/F., C&C Building, 36 Ting Lai Road, Tai Po, N.T., Hong Kong
電話 Tel 2150 2100
傳真 Fax 2407 3062
電郵 Email info@suplogistics.com.hk

承印者 Printer
合群(中國)印刷包裝有限公司
Powerful (China) Printing & Packing Co., Ltd.

出版日期 Publishing Date
二OO九年三月第一次印刷 First print in March 2009
二O一七年一月第六次印刷 Sixth print in January 2017

Published in Hong Kong by Forms Kitchen,
a division of Wan Li Book Company Limited.
ISBN 978-988-8137-08-4